SECTION 105
PERMITS

SECTION 105
PERMITS

SECTION 308
PIPING SUPPORT

SECTION 312
TESTS

CHAPTER 4
FIXTURES

SECTION 405
INSTALL FIXTURES

SECTION 412
FAUCETS & FITTINGS

CHAPTER 5
WATER HEATERS

CHAPTER 6
WATER SUPPLY & DISTRIB

SECTION 606
WATER DISTRIB INSTALL

SECTION 606
WATER DISTRIB INSTALL

TABLE 608.16.1
MIN AIR GAPS

SECTION 704
DRAIN PIPE INSTALL

SECTION 708
CLEANOUTS

TABLE 709.1
DFU TABLE

CHAPTER 8
INDIRECT WASTE

SECTION 918
AIR ADMIT VALVES

CHAPTER 14
SUBSURFACE SYSTEMS

CHAPTER 15
REF STANDARDS

APPENDIX A
FEE SCHEDULE

APPENDIX A
FEE SCHEDULE

APPENDIX B
RAINFALL RATES

APPENDIX B
RAINFALL RATES

APPENDIX C
STRUCTURAL SAFETY

APPENDIX C
STRUCTURAL SAFETY

APPENDIX D
DESIGN TEMPS

APPENDIX D
DESIGN TEMPS

APPENDIX E
WATER PIPE SIZING

APPENDIX F
BOARD OF APPEALS

APPENDIX F
BOARD OF APPEALS

RESOURCE A
PLUMBING PROV